John Milne Bramwell

Suggestion

Its Place in Medicine and Scientific Research

John Milne Bramwell

Suggestion
Its Place in Medicine and Scientific Research

ISBN/EAN: 9783337371456

Printed in Europe, USA, Canada, Australia, Japan

Cover: Foto ©berggeist007 / pixelio.de

More available books at **www.hansebooks.com**

SUGGESTION:

ITS PLACE IN MEDICINE AND SCIENTIFIC RESEARCH.

Being a Lecture delivered on behalf of the

LEIGH BROWNE TRUST,

At St. Martin's Town Hall, on February 9th, 1897,

BY

Dr. J. MILNE BRAMWELL.

SUGGESTION: ITS PLACE IN MEDICINE AND SCIENTIFIC RESEARCH.

SUGGESTION has ever played an important part in medicine. In earlier and more superstitious times the priest or saint was the physician; suggestion was administered in concrete form through the medium of saintly relics, or holy wells, and the cure was ascribed to Divine agency. As superstition decreased, belief in the curative power of saintly relics diminished, and the cures which were said to have been wrought by their means were usually looked upon as idle tales. Still later, science pointed out how every function in the human body could be influenced by fear, hope, and other emotional states; then the cures we are referring to were admitted to be possible, while the saintly relics were regarded simply as the means by which the emotional states were evoked.

In 1839, as the result of the researches of
Sir Henry Holland, the action of the mind
upon the body was still further realised.   He
drew attention to the fact that, though the
influence of the emotions upon physical
conditions had long been the subject of study,
the effects of the consciousness, directed by
distinct voluntary effort to particular parts of
the organism, had been almost entirely over-
looked.   In his opinion many of the functions,
and all the sensations of the human body,
could be influenced by voluntarily fixing the
attention upon some function or organ, even
when this was unattended by emotion or
anxiety.

Sir Henry Holland's theory was quickly
seized upon by Braid, who thought he had
found in it, not only an explanation of the
action of certain drugs, but also of the
phenomena of hypnotism.   According to
Braid, the action of homœopathic remedies,
when these were given in attenuated solutions,
was a purely subjective one ; he also believed
that the mental element associated with the
administration of drugs in general had been
far too much ignored.   Braid considered that

all medicines had a two-fold action ; one due to their physical properties, the other to the patient's expectation of a definite result.

Many authorities still believe that suggestion is largely intermixed with ordinary medical practice, and often forms the most important factor in its success. According to Dr. Wilks: " To sit down in one's chair and write on a piece of paper the name of some drug for every ailment, without exception, which comes under our observation, is, in the present state of medicine, an absurdity, and is simply a pandering to human weakness. I do not say that drugs are not useful in a moral sense. I am merely contending that the method is not scientific, as we usually apply this term. I know of no more successful practitioner than the late Sir William Gull, and his treatment was rational; but he did not credit any particular drug with the properties ascribed to it by the patient. His prescriptions very often consisted of nothing but coloured water."

Dr. Wilks states that changes in the pathological views of disease have caused the whole method of its treatment to be altered again

and again, and that further chemical know-
ledge frequently shows that the drugs we
employ do not possess the qualities we have
been in the habit of attributing to them.

The late Dr. Hack Tuke, in speaking of
Sir Andrew Clarke, said : "His favourite drugs
were bi-carbonate of potash and a vegetable
bitter, but neither drugs nor diet formed the
central factor of his treatment, or explained
its success. 'Suggestion' lay at the root of
it all. The term, however, is too mild, unless
understood in the technical sense in which it
has been employed in recent times. In short,
Sir Andrew out-Bernheimed Bernheim ; he
was, in a word, the most successful hypnotist
of his day."

The theory, held by Professor Benedikt, of
Vienna, that magnets possess extraordinary
therapeutic powers, is an interesting modern
example of the unconscious use of suggestion
in medicine. According to Benedikt, certain
forms of hysteria are better treated by the
magnet than by electricity, hydropathy, or
drugs. When the magnet is applied to the
sensitive vertebræ, without removal of the
dress, the irritable patient soon becomes quiet,

or even quasi-paralysed. The muscles gradually relax, the respiration becomes sighing, consciousness slowly disappears. The resistance to conduction in motor nerves could easily become absolute. The two poles have different effects. The magnet must be applied with caution; the patients may be injured by it.

These statements were recently tested in America ; electro-magnets of enormous power were used (2,000 to 5,000 C.G.S. units to the square centimetre), and experiments were made on human subjects and lower animals. A young dog was subjected to magnetic influence for five hours (apparently an absolutely painless experiment); but, instead of being paralysed from the increased resistance to conduction in motor nerves, on being liberated it was more lively than before. The experimenters conclude that the human organism is in no wise appreciably affected by the most powerful magnet known to modern science ; that neither direct nor reversed magnetism exerts any perceptible influence upon the iron contained in the blood, upon the circulation, upon ciliary or protoplasmic movements, upon sensory or motor nerves, or upon the brain. The ordinary magnets used in medicine, they say, have a purely suggestive or psychic effect, and would, in all probability, be just as useful if made of wood.

Admitting that Dr. Wilks' views as to the value of drugs are distinctly pessimistic, we must still grant the importance of mental influences in relation to the ordinary medical treatment of disease. On this point, however, I shall not dwell further, but will pass at once to the subject of hypnotism.

Modern hypnotism undoubtedly owes its origin to mesmerism, and to understand its evolution, a clear conception of mesmeric theories is necessary ; of these the views of Esdaile may be regarded as a fair summary.

JAMES ESDAILE ; HIS MESMERIC THEORY AND PRACTICE.

James Esdaile was born at Montrose in 1808. He studied medicine at Edinburgh University, graduated there as M.D. in 1830 ; obtained a medical appointment in the service of the East India Company, and arrived at Calcutta in July, 1831.

Esdaile made his first mesmeric experiment in 1845, when in charge of the Native Hospital at Hooghly. The subject was a Hindoo convict who had just undergone a painful surgical operation, which was about to be repeated. At this time, Esdaile knew nothing of mesmerism, except what he had read in the daily papers, but it occurred to him to try to mesmerise this patient in order to render him insensible to pain. The experiment was successful, and was again repeated on the same subject a week later. This encouraged Esdaile to persevere, and at the end of a year he reported 120 painless operations to the Government. As up to this date no anæsthetic of any kind had been employed in surgical operations in India, his cases naturally excited much attention. A committee, largely composed of medical men, was appointed to investigate his work; their report was favourable, and Esdaile was placed at the head of a Government Hospital in Calcutta, for the express purpose of mesmeric practice. From this date, until he left India in 1851, he occupied similar posts. He recorded 261 painless capital operations and

many thousand minor ones, and reduced the
mortality in the removal of the enormous
tumours of elephantiasis from 50 to 5 per
cent.   Patients flocked to him from all parts
of the country, and the record of his painless
mesmeric operations forms one of the most
fascinating and romantic pages in the whole
history of medical science.

Esdaile believed that mesmeric phenomena
were due to the action of a vital curative
fluid, or peculiar physical force, which, under
certain circumstances, could be transmitted
from one animal to another.   Various inanimate objects, such as metals, crystals and
magnets, were also supposed to possess this
force or fluid, and to be capable of inducing
and terminating the mesmeric state, and
of exciting, arresting and modifying its
phenomena.   One metal, for example, would
produce catalepsy, another change this into
paralysis ; a glass of water could be changed
with odyllic force by being breathed upon by
the mesmeriser.   Esdaile thus summarises his
theory of the therapeutic action of mesmerism :
" There is good reason to believe that the
vital fluid of one person can be poured into

the system of another. A merciful God has engrafted a communicable, life-giving, curative power in the human body, in order that when two individuals are found together, deprived of the aids of art, the one in health may often be able to relieve his sick companion, by imparting to him a portion of his vitality."

After leaving India, in 1851, Esdaile settled in Perth, and about a year later informed Elliotson that he had found the inhabitants of the far North as susceptible to mesmerism as those of the farest East. Dr. Fraser Thompson, Physician to the Perth Infirmary, became a convert and performed some successful mesmeric operations. His colleagues, however, called a meeting of the Directors, and stated that they would resign if the practice of mesmerism were permitted in the Hospital.

Esdaile's last days were undoubtedly embittered by the mesmeric controversy raging in England, of which the following case affords an illustration:—Mr. Ward, a Nottinghamshire surgeon, amputated a thigh during mesmeric trance; the patient did not make a single muscular movement during the entire operation, and was afterwards unable to recall the slightest sensation of pain. The case was reported to the Royal Medical and Chirurgical Society, but was badly received. Dr. Copland proposed that no account of such a paper having been read before the Society should be entered in its minutes ; for pain was a wise provision of Nature with which it was impious to interfere. He considered that patients ought to suffer pain while their surgeons were operating upon them, as this did them good and enabled them to recover better.

James Braid, the Father of Hypnotism.

James Braid commenced to investigate the subject of mesmerism in 1841. He was born in Fifeshire in 1795, had studied at Edinburgh University, and at this time was in practice in Manchester, where he had already gained a high reputation as a skilful surgeon. Braid believed mesmeric phenomena were due to self-deception or trickery, and, at the first mesmeric *séance* at which he assisted, saw nothing to cause him to alter his views. Six days later he noticed that one subject was unable to open his eyes. He regarded this as a real phenomenon and was anxious to discover its physiological cause; and the following evening, when the case was again operated on, he believed he had done so. After making a series of experiments, chiefly on personal friends and relatives, he expressed his conviction that the phenomena he had witnessed were purely subjective, and commenced almost immediately to place these views before the public. In 1843 Braid published "Neurypnology, or the Rationale of Nervous Sleep." This was followed by many other works of more or less importance, and of these I have been able to trace 32, but all have long been out of print in this country.

According to Braid, the phenomena of mesmerism depended entirely on the physical and psychical condition of the patient, and were absolutely independent of the volition of the operator, or of any mystical or magnetic fluid which emanated from him. From the physiological side he explained the phenomena by changes in the nervous system of the subject. These consisted in the exhaustion of certain nerve centres, with resulting decrease in the functional activity of the central nervous system ; they arose from continued monotonous stimulation of other nerves, *e.g.*, those of the eye by fixed gazing ; those of the skin by passes with contact.

He explained the phenomena psychologically by concentration of attention and monoideism. The mind was so engrossed with a single idea as to render it dead to all other influences ; the attention was concentrated upon the particular function called into action, while the others passed into a state of torpor. Only one function was active at any one time, and hence intensely so ; the arousing of any dormant function was equivalent to superseding the one in action.

Braid proposed to substitute the term hypnotism for that of mesmerism, and invented the general terminology of the subject, which remains little altered to the present day. He performed many experiments in order to test the alleged powers of magnets, metals, drugs in sealed tubes, &c., and found that the phenomena described by the mesmerists appeared when the patients had preconceived ideas on the subject, or when these were excited by leading questions, but were invariably absent when they were ignorant of what was being done. Real magnets had no effect when the patients were unaware of their presence, while pretended magnets produced the phenomena when the patients knew what was expected to occur ; and thus the mind of the patient alone was sufficient to produce the effects attributed to magnetic or odyllic force. Many cases of alleged clairvoyance and thought transference were also investigated by Braid; but he was never able to find anything but hypnotic exaggeration of natural powers.

DANGERS.

According to Braid, the hypnotic subject acquired new and varied powers, but did not at the same time lose his volition or moral sense. During hypnosis the patients evinced great docility, but were quite as fastidious of correct conduct as when in the natural state ; they would neither reveal secrets nor accept improper suggestions. Braid stated that he had proved that no one could be affected by hypnotism at any stage unless by voluntary compliance.

MEDICAL CASES.

Braid recorded numerous medical cases which were relieved or cured by hypnotism. The majority of these were functional nervous disorders, but remarkable results were also obtained in many cases of organic disease. Of the latter, the clearing up of a long-standing corneal opacity, in a patient who was being hypnotised for the relief of a rheumatic affection, is an interesting example. Marked improvement was also obtained in many cases of organic paralysis.

Hypnosis, in Braid's opinion, was not necessarily associated with loss of consciousness, and in many of his most successful cases the patients were afterwards able to recall all that had taken place. He claimed that he could hypnotise his patients more quickly than the mesmerists could influence theirs, and also that his curative results were superior, despite the fact that he neither believed in, nor invoked, occult powers.

In 1859, Dr. Azam of Bordeaux became acquainted with Braid's hypnotic work and commenced to investigate the subject for himself; an account of his experiments, with much reference to Braid, appeared in the Archives de Médecine in 1860. From this date, the subject of hypnotism was never lost sight of in France, but it was not until forty years after its original publication that " Neurypnology " was translated by Dr. Jules Simon.

Dr. A. A. Liébeault.

Liébeault was born in 1823, and commenced to study medicine in 1844. In 1848, he read a book on animal magnetism ; this impressed him greatly, and a few days later he successfully mesmerised several persons. He received his M.D. in 1850, and shortly afterwards started country practice. He worked hard, and was often in the saddle making his rounds at 2 a.m. He had no private fortune, but in ten years he saved enough to enable him to live independently of his profession. In 1860, he began to study mesmerism seriously, just at the time that Velpeau communicated Azam's experiments to the Académie de Médecine. In order to find subjects for experiment, Liébeault took advantage of the parsimonious

character of the French peasant. His
patients had absolute confidence in him,
but they had been accustomed to be
treated in the ordinary manner. He, therefore, said to them : "If you wish me to treat
you with drugs, I will do so, but you
will have ·to pay me as formerly. On the
other hand, if you will allow me to hypnotise
you, I will do it for nothing." He soon had
so many patients that he was unable to find
time for necessary repose or study. In 1864,
he settled in Nancy; lived quietly on the
interest of his capital, and practised hypnotism
gratuitously among the poor.

For two years he worked hard at his
book, "Du Sommeil et des États analogues,
considérés surtout au point de vue de l'action de
la morale sur le physique," but of this one copy
alone was sold. His colleagues regarded him
as a madman ; the poor as their Providence,
calling him "the good father Liébeault."
His clinique was crowded with patients ;
of these he cured many who had vainly sought
help elsewhere, and few left him without
having received benefit. In 1882, Liébeault
cured an obstinate case of sciatica of six

years' duration, which Bernheim had treated in vain for six months. In consequence of this Bernheim visited Liébeault. This was a great event in the life of the humble doctor. At first Bernheim was sceptical and incredulous, but soon this changed into admiration. He multiplied his visits, and became a zealous pupil and true friend of Liébeault. In 1884, Bernheim published the first part of his book, " De la Suggestion," which he completed in June, 1886, by a second part, entitled " La Thérapeutique suggestive." From this date, Liébeault's name became known throughout all the world. The first edition of his book was quickly bought up, and doctors flocked from all countries to study the new therapeutic method.

In the summer of 1889, I spent a fortnight at Nancy in order to see Liébeault's hypnotic work. His clinique, invariably thronged, was held in two rooms situated in a corner of his garden. The interior of these presented nothing likely to attract attention ; and, indeed, anyone coming with preconceived ideas of the wonders of hypnotism would be greatly disappointed. For, putting aside the methods of treatment, and some slight differences probably due to race-characteristics, one could easily have imagined oneself in the out-patient department of a general hospital. The patients perhaps chatted more freely amongst themselves, and questioned the doctor in a more familiar way than one had been accustomed to see in England. They were taken in turn, and the clinical casebook referred to ; hypnosis was then rapidly induced by the method about to be described, suggestions given, and notes taken, the doctor maintaining the while a running commentary for my benefit.

The patient was placed in an armchair, told to think of nothing, and to look steadily at the operator. This fixation was not maintained long enough to produce any fatigue of the eyes, and appeared to be simply an artifice for arresting the attention. If the eyes did not close spontaneously, Liébeault requested the patient to shut them, and then proceeded to make the following suggestions, or others resembling them :—" Your eyelids are getting heavy, your limbs feel numb, you are becoming more and more drowsy, &c."

Nearly all the patients I saw were easily and rapidly hypnotised, but Liébeault informed me that nervous and hysterical patients were his most refractory subjects.

As I was a stranger, an exception was made in my favour, and I was shown a few hypnotic experiments; but cure, not experiment, seemed the sole object. The quiet ordinary everyday tone of the whole performance formed a marked contrast to the picture drawn by Binet and Féré of the morbid excitement shown at the Salpêtrière. The patients told to go to sleep, apparently fell at once into a quiet slumber, then received their dose of curative suggestions, and when told to awake, either walked quietly away or sat for a little to chat with their friends; the whole process rarely lasting longer than ten minutes. The negation of all morbid symptoms was suggested; also the maintenance of the conditions upon which general health depends, *i.e.*, sleep, digestion, &c. I noticed that in some instances curative suggestions appeared to be perfectly successful, even when the state produced was only that of somnolence. The cases varied widely, and most of them were either relieved or cured. No drugs were given; and Liébeault took especial pains to explain to his patients that he neither exercised nor possessed any mysterious power, and that all he did was simple and capable of scientific explanation.

Two little incidents, illustrating the absence of all fear in connection with Liébeault and hypnotism, interested me greatly.

A little girl about five years old, dressed shabbily, but evidently in her best, with a crown of paper laurel leaves on her head and carrying a little book in her hand, toddled into the sanctum, fearlessly interrupted the doctor in the midst of his work, by pulling his coat, and said : " You promised me a penny if I got a prize." This, accompanied by kindly words, was smilingly given, incitement to work having been evoked in a pleasing, if not scientific way. Two little girls, about six or seven years of age, no doubt brought in the first instance by friends, walked in and sat down on a sofa behind the doctor. He, stopping for a moment in his work, made a pass in the direction of one of them and said : " Sleep, my little kitten," repeated the same for the other, and in an instant they were both asleep. He rapidly gave them their dose of suggestion and then evidently forgot all about them. In about twenty minutes one awoke, and wishing to go essayed, by shaking and pulling, to awaken her companion—her amused expression of face, when she failed to do so being very comic. In about five minutes more the second one awoke, and hand in hand they trotted laughingly away.

Braid anticipated many of the most important observations of the School of Nancy ; but we ought not, on that account, to undervalue the services of that School, and more especially those of its founder—Liébeault. Braid's researches were undoubtedly the exciting cause of the hypnotic revival in France, but little or nothing was known of any of his works except "Neurypnology," and his last MS., which contained some of his later views, was not published in France until 1883. Liébeault independently arrived at the conclusion that the phenomena of hypnotism were purely subjective in their origin, and to him we owe the development of modern hypnotism.

Another point in reference to their career is worthy of note. Braid's views at once brought him fame. His books sold rapidly, the demand for them exceeding his power of supply. The medical journals were open to him, to an extent which may well excite envy in those interested in the subject at the present day. Liébeault's book, on the contrary, remained unsold ; his statements only found sceptics, his methods of treatment were rejected without examination, and he was laughed at and despised by all. From the day he settled in Nancy in 1864, until Bernheim—some twenty years later— was the means of bringing him into notice, Liébeault devoted himself entirely to the poor, and refused to accept a fee lest he should be regarded as attempting to make money by unrecognised methods. Even in his later days, fortune never came to him, nor did he seek it, and his services—services, which he himself with true modesty described as the contribution of a single brick to the edifice many were trying to build—only began to be appreciated when old age compelled him to retire from active work. Though his researches have been recognised, it is certain that they have not been estimated at their true value, and that members of a younger generation have reaped the reward which his devotion of a lifetime failed to obtain.

The term "School of Nancy" has been applied to Liébeault and his colleagues; but, as Professor Beaunis points out, they do not claim to have originated a School, and, though they agree on certain points, differ widely on others.

According to Liébeault, and other members of the so-called School of Nancy, hypnosis is a physiological condition which can be induced in those who enjoy perfect health, and its phenomena are analogous to those of normal waking and sleeping life. Everybody, they say, is liable to be influenced by suggestion; this susceptibility is increased in hypnosis and forms the sole distinction between it and the normal state.

To this view several objections might be raised. The success of suggestion depends not on the suggestion itself but on conditions inherent in the subject. These are (1) the willingness to accept and carry out the suggestion, and (2) the power to do so. In the hypnotised subject, except in reference to criminal or improper suggestions, the first condition is generally present. The second varies according to the depth of the hypnosis and the personality of the patient. For instance, I might suggest analgesia, in precisely similar terms, to three subjects and yet obtain quite different results. One might become profoundly analgesic, the second slightly so, and the third not at all. Just in the same way, if three jockeys attempt to make their horses gallop a certain distance in a given time, the suggestions conveyed by voice, spur, and whip may be similar, and yet the results quite different. One horse, in response to suggestion, may easily cover the required distance in the allotted time ; it was both able and willing to perform the feat. The second, in response to somewhat increased suggestion, may nearly do so ; it was willing, but had not sufficient staying power. The third, able but unwilling, not only refuses to begin the race, but bolts off in the opposite direction. With this horse we have the exact opposite of the result obtained in the first instance ; and yet possibly the amount of suggestion it received largely exceeded that administered to the others.

As Myers has pointed out, the operator directs the conditions upon which hypnotic phenomena depend, but does not create them. " Professor Bernheim's command, 'Feel pain no more,' is no more a scientific instruction *how* not to feel pain, than the prophet's ' Wash in Jordan and be clean ' was a pharmacopœial prescription for leprosy." In hypnosis, the essential condition is not the means used to excite the phenomena, but the peculiar state which enables them to be evoked.

Suggestion no more explains the phenomena of hypnotism than the crack of a pistol explains a boat race. Both are simply signals—mere points of departure, and nothing more. In Bernheim's hands the word suggestion has acquired an entirely new signification, and differs only in name from the " odyllic " force of the mesmerists. It has become mysterious and all-powerful, and is supposed to be capable, not only of evoking and explaining all the phenomena of hypnotism, but also of origina- ting—nay, even of being—the condition itself. According to this view, suggestion not only starts the race, but also creates the rowers and builds the boat !

While Liébeault, Bernheim and others believe that the hypnotic state is practically identical with the normal, they at the same time hold that the volition is weakened or suspended in hypnosis, and that disagreeable, or even criminal acts, can sometimes be successfully suggested to the subject. This question I will discuss later in connection with another theory.

CHARCOT'S THEORY, OR THAT OF THE SALPÊTRIÈRE SCHOOL.

According to this School, hypnosis is an artificially induced morbid condition — a neurosis only to be found in the hysterical, and its phenomena can be produced by purely physical means. This theory cannot be accepted unquestioned, for, as the following statistics show, if the hysterical alone can be hypnotised, then over 90 per cent. of mankind must suffer from hysteria. Some years ago, Bernheim had already attempted to hypnotise 10,000 hospital patients, with over 90 per cent. of successes, while Dr. Wetterstrand, of

Stockholm, recently reported 6,500 cases, with 3 per cent. failures. Schrenck-Notzing, in his First International Statistics, published in 1892, gave 8,705 cases by 15 observers in different countries, with 6 per cent. of failures. Mr. Hugh Wingfield, when Demonstrator of Physiology at Cambridge, attempted to hypnotise over 170 men, all of whom, with the exception of 18, were undergraduates. In about 80 per cent. hypnosis was induced at the first attempt ; but, as no second trial was ever made with the unsuccessful cases, these results undoubtedly understate the susceptibility.

Most of the undergraduates would be drawn from our public schools ; and, if these do not always turn out good scholars, they cannot at least be accused of producing hysterical invalids. Braid stated that the nervous and hysterical were the most difficult to hypnotise, while Liébeault found soldiers and sailors particularly easy to influence. Grossmann, of Berlin, recently asserted that hard-headed North Germans were very susceptible, and I observed that healthy Yorkshire farm labourers made remarkably good subjects. Professor Forel told me that.

he had hypnotised nearly all his asylum warders; that he selected these himself, and certainly did not choose them from the ranks of the hysterical.

These and similar facts apparently justify the statements of Forel and Moll that it is not the healthy, but the hysterical who are the most difficult to hypnotise. According to the former, "every mentally healthy man is naturally hypnotisable;" while the latter says: "If we take a pathological condition of the organism as necessary for hypnosis, we shall be obliged to conclude that nearly everyone is not quite right in the head. The mentally unsound, particularly idiots, are much more difficult to hypnotise than the healthy. Intelligent people, and those with strong wills, are more easily hypnotisable than the dull, the stupid, or the weak willed."

CAN VARIOUS PHYSICAL AND MENTAL PHENO-
MENA BE EXCITED BY THE APPLICATION
OR NEAR PRESENCE OF CERTAIN METALS,
MAGNETS, AND OTHER INANIMATE OBJECTS?

Here, in the assertions of the Salpêtrière School and their refutation by that of Nancy, we have an exact counterpart of the controversy between Braid and the mesmerists. All the old errors, the result of ignoring mental influences, are once more revived. Medicines are again alleged to exercise an influence from within sealed tubes. The physical and mental conditions of one subject are stated to be transferable to another, or even to an inanimate object. It is useless to enter into any arguments to refute these statements, for this would be needlessly repeating the work of Braid. Indeed, in many instances, their absurdity renders argument unnecessary. For example, when a sealed tube, containing laurel-flower water was brought near a Jewish prostitute, she adored the Virgin Mary! The chief apostle of these doctrines is Luys ; and considerable attention was drawn to them in this country in 1893 by popular articles in the daily papers and elsewhere. Indeed, the editor of a well-known medical journal thought them of sufficient importance to demand his writing a book in order to disprove them. He apparently was ignorant of the fact that M. Dujardin-Beaumetz had, in 1888, reported to the Académie de Médecine that Luys' experiments were conducted so carelessly as to rob them of all value, and that among students of hypnotism they are entirely disregarded.

THE SUBLIMINAL CONSCIOUSNESS THEORY.

Within recent times another theory has arisen. This, instead of explaining hypnotism by the arrested action of some of the brain centres which subserve normal life, attempts to do so through the arousing of certain powers over which we normally have little or no control. This theory appears under various names, "Double Consciousness," "Das Doppel-Ich," &c., and the principle on which it depends is largely admitted by science. William James, for example, says: "In certain persons, at least, the total possible consciousness may be split into parts which co-exist, but mutually ignore each other."

The clearest statement of this view is given by F. W. H. Myers ; he suggests that the stream of consciousness in which we habitually live is not our only one. Possibly our habitual consciousness may be a mere selection from a multitude of thoughts and sensations, some at least equally conscious with those we

empirically know.   No primacy is granted by
this theory to the ordinary waking self, except
that among potential selves it appears the
fittest to meet the needs of common life.   As
a rule, the waking life is remembered in
hypnosis, but the hypnotic life is forgotten
in the waking state ; this destroys any claim
of the primary memory to be the sole memory.
The self below the threshold of ordinary
consciousness Myers terms the " subliminal
consciousness," and the empirical self of
common experience the " supraliminal."   He
holds that to the subliminal consciousness and
memory a far wider range, both of physiological
and psychical activity, is open than to the
supraliminal.   The subliminal, or hypnotic
self, can exercise over the nervous, vasomotor
and circulatory systems a degree of control
unparalleled in waking life.

According to the late Professor Delboeuf,
the hypnotic subject's power of regulating
his own organism may be a revival of that
possessed by lower animal types—the possible
remote ancestors of the human race.   In the
latter, he said, the animal was just as conscious of what was taking place in its interior

as it was of what was happening at its periphery. With the progress of development, however, the care of the vegetative life has been handed over by the will to nervous mechanisms which have learnt to regulate themselves, and which in general fulfil their task to perfection. Sometimes the machine goes wrong, and intervention becomes desirable. The power which formerly voluntarily regulated it has, however, dropped out of the normal consciousness, and if we desire to find a substitute for it we must turn to hypnotism.

In the hypnotic state the mind is in part drawn aside from the life of relation, while at the same time it preserves its activity and power. Voluntary attention can be abstracted from the outer world and directed with full force upon a single point, and the hypnotic consciousness is thus able to put in movement machinery which the normal consciousness has lost sight of and ceased to regulate. It may be able to act, not only on the reflexes, but on the vasomotor system, on the unstriped muscles, on the apparatus of secretion, &c. If a contrary opinion has till now prevailed, this is because observation has been exclusively directed to the normal exercise of the will. It can, however, in the hypnotic state, regulate movements which have become irregular and assist in the repair of organic injuries. In a word, hypnotism does not depress but exalts the will, by permitting it to concentrate itself upon the point where disorder is threatened.

By this theory Delboeuf attempts to explain the mechanism of the inverse action of the moral on the physical, which is sometimes, in his opinion, almost, if not quite, equal to that of the physical on the moral.

The following is one of the most interesting of the experiments upon which he formed his conclusions. The subject, J., was a healthy young woman, who had for several years been one of his servants, and whom he had previously hypnotised. Delboeuf first explained to her what he wished to do, and obtained her consent in the waking state. He then extended both her arms upon a table, heated red-hot a bar of iron, eight millimetres in diameter, and applied it to both of them, taking care that the burns should be identical in duration and extent; while at the same time he suggested that she should feel pain in the left arm alone.

The operation was performed at seven o'clock in the evening, and immediately afterwards each arm was covered with a bandage. During the night J. had pain in the left arm, but felt nothing in the right. Next morning Delboeuf removed the bandages; the right arm presented a defined eschar, the exact size of the iron, without inflammation or redness; on the left was a wound of about three centimetres in diameter with inflamed blisters. Next day the left arm was much worse, and J. complained of acute pain. Delboeuf then hypnotised her, and removed the pain by suggestion. The wound dried and inflammation rapidly disappeared.

In Delboeuf's opinion the persistent belief that one is suffering from disease may ultimately cause disease, and, in the same way, the conviction that a morbid condition does not exist may contribute to its disappearance. He considers that the organic changes, which follow such an injury as we have just described in the case of J., are not alone due to the injury itself, but are also partly caused by the patient's consciousness of pain. The absence or presence of pain may, to a greater or lesser extent, influence vasomotor conditions. On the one hand, organic injury, unassociated with pain, may not be followed by congestion, inflammation or suppuration, while in an identical injury, accompanied by pain, these conditions may be present. The consciousness of pain, in addition to being sometimes responsible for morbid changes at the site of injury, may also help to spread them to other parts more or less remote, and thus, when pain is removed or relieved, this really means the disappearance or decrease of one of the factors in the organic malady.

This freedom from pain under conditions of nerve and tissue with which it is usually inevitably connected, Myers regards as one of the great dissociative triumphs of hypnotism. Some intelligence, he says, is involved in a suppression thus achieved; for this is obtained, not, as with narcotics, by a general loss of consciousness, but by the selection and inhibition from among all the percipient's possible sensations of disagreeable ones alone. This is not a mere anæsthetisation of some particular group of nerve-endings, such as cocaine produces; it involves the removal also of a number of concomitant feelings of nausea, exhaustion, anxiety, not always directly dependent on the principal pain, but needing, as it were, to be first subjectively distinguished as disagreeable before they are picked out for inhibition.

HYSTERIA : A DISEASE OF THE SUBLIMINAL SELF.

Myers does not consider the subliminal self free from disease any more than the supraliminal ; subliminal disturbances may arise and make themselves felt in the supraliminal being. He draws attention to the analogy which exists between the changes in the nervous vasomotor and circulatory systems which occur in hypnosis, and those presented by hysteria. Hysterical phenomena, in his view, are produced by self-suggestions of an irrational and hurtful kind; they are diseases of the hypnotic substratum. Hypnotism is not a morbid state ; it is the manifestation of a group of perfectly normal, but habitually subjacent powers, whose beneficent operation we see in cures by therapeutic suggestion, whose neutral operation we see in ordinary hypnotic experiment, and whose diseased operation we see in the vast variety of self-suggested maladies.

The following are the points most worthy of note in the theories of Myers and Delboeuf:—

1. The chief characteristic of the hypnotic state is the subject's far-reaching power over his own organism.—All observers, from Braid onwards, have recognised that the mental and physical conditions of the hypnotic subject could be influenced to an extent and in a fashion unparalleled in waking life. Of this many examples are to be found in the records of psychological and physiological experiments, and in numerous cases of disease relieved or cured by hypnotic treatment. Myers and Delboeuf, however, were apparently the first since Braid to realise that it is the subject, not the operator, who controls these powers.

2. The hypnotic powers are exercised intelligently by the subject, and manifest an increased, not diminished, volition.—Both Myers and Delboeuf draw attention to the fact that the volition is unimpaired in hypnosis. The former cites instances in which the moral sense is apparently increased, and

the latter states that hypnosis does not depress, but exalts the will, by permitting it to concentrate itself upon the point where disorder is threatened.   Further, Myers shows that the inhibition of pain in hypnosis involves several intelligent and complicated acts.

3. Subliminal or subconscious states are more clearly defined.—Many cases of alternating consciousness have been observed in the non-hypnotised subject. As a rule this has been associated with hysteria, or some other morbid condition. Sometimes the primary waking state has been morbid, the secondary one comparatively healthy.   Of this class, Félida X., so ably described by Dr. Azam, is the familiar example.

The works and writings of Edmund Gurney, A. T. Myers, Frederic Myers, Pierre Janet, William James, and many others, have rendered us familiar with the phenomenon of secondary or multiple consciousness in hypnosis. It can be experimentally demonstrated not only that the hypnotised subject possesses a secondary consciousness, which alternates with his primary one, but also that it is possible for the two to co-exist and to

manifest different phenomena simultaneously.
For example, an individual may have his
attention concentrated upon the act of reading
aloud from a book with which he was previously unacquainted, and, at the same instant,
hè may be writing automatically—as far as
his primary consciousness is concerned—the
result of a problem, suggested to him in
hypnosis the moment before that state was
terminated. The primary waking consciousness
retains no recollection of the hypnotic suggestion ; does not know that the secondary
consciousness, after the hypnotic state has
been terminated, first solves the problem and
then directs the motor acts which record it ;
and is also unconscious of the motor acts
themselves.

4. Hysteria is a disease of the hypnotic
substratum of the personality.—Myers' theory
that hysteria is a disease of the hypnotic
substratum is an extremely ingenious one,
and is the only reasonable explanation of
the resemblance between certain hypnotic
and hysterical phenomena with which I
am acquainted. As we have seen, those
who believed that hypnosis and hysteria

were identical stated that the hysterical alone could be hypnotised. On the other hand, those with wider experience have successfully demonstrated that the hysterical are generally, if not invariably, the most difficult to influence. Of this fact Myers' theory possibly affords an explanation. May not the difficulty of inducing hypnosis in the hysterical—of making one's suggestions find a resting-place in them—be due to the fact that the hypnotic substratum of their personality is already occupied by irrational self-suggestions which their waking will cannot control ?

5. The explanation of hypnotic powers by a supposed revival of those formerly possessed by lower non-human ancestral types.—Granting that one or more subconscious states exist in the human personality, and that hypnotic phenomena owe their origin to the fact that we have by some means or other succeeded in tapping them, two questions still remain :—

Thus, let it be supposed that I possess a friend called Brown, who is usually, physically and mentally, an ordinary individual ; from time to time, however, he manifests an extra-ordinary increase of physical power.   Again, though still more rarely, he displays a range of mental powers of which he had formerly given no indication. I ask for an explanation. I am told that Brown, as I know him best, is indeed Brown; but that his increased physical powers are due to the fact that when he shows them he is Jones, and his increased mental ones to the further fact that he is then Robinson.   Granting that the phenomena afford evidence of three separate personalities, I cannot accept this explanation as a solution of the problem in its entirety.   I want to know first, how did Jones and Robinson acquire their powers, and secondly, what has been done to or by Brown which has enabled these powers to be evoked ?

A. How did Jones and Robinson acquire their powers ?

Is it reasonable to suppose, with Delboeuf, that the hypnotic powers, regarded as a whole, existed in some ancestral type ?   Granting that a limited analogy exists between lower forms of animal life and hypnotised subjects as to their power of influencing certain physical conditions, it would, I think, be impossible to establish an analogy between the mental and moral powers of the latter and those of the savage or lower animal. For example, one of my patients suddenly developed arithmetical powers far exceeding those she possessed in the normal state. She is not likely to have derived them from some savage ancestor, who was unable to count beyond five, or from some lower animal, presumably ignorant of arithmetic. Again, the same patient spontaneously solved in hypnosis a difficult problem in dressmaking. The power of correctly designing a garment, in accordance with the passing fashion of the present day, can hardly have been derived from some woad-stained ancestor, or lower animal form. Further, the increased modesty of the hypnotised subject, his greater power of controlling or checking morbid passions or cravings does not find its counterpart in the savage or ape.

B. What is the connection between hypnotic methods and the production of hypnotic phenomena ?

To this, I think, no reasonable answer has been given. Personally, I can see no logical connection between the acts of fixed gazing, concentration of attention, or of suggested ideas of drowsy states, and the wide and varied phenomena of hypnosis. Hypnotic phenomena do not appear spontaneously, and one or other of the methods described must have been employed in each case before hypnosis was induced in the first instance. But I cannot conceive the idea that the methods explain the phenomena.

VOLITION IN HYPNOSIS. SUGGESTED CRIMES.

In direct opposition to the views of Braid, the writings of the Nancy School indicate a belief that the hypnotic state is essentially characterised by the obedience of the subject to the operator. Some years ago much stress was laid upon complete obedience; but now, possibly as the result of the influence of Professor Delboeuf, a greater power of resistance is conceded. The principal objection to the statement that certain hypnotic subjects will accept improper or criminal suggestions is to be found in the fact that it is based solely on laboratory experiments, and not on the observations of actual crime or impropriety.

The following is a typical case. It is suggested to a high-principled girl, in the alert stage of somnambulism with her eyes open, that she shall take a lump of sugar from the basin and put it into her mother's teacup, after having been informed that it is really a piece of arsenic certain to cause death. Bernheim alleges that the subject believes she has committed a real crime; but it is necessary to examine the facts upon which this grave statement is founded. All that the somnambule did was to put a piece of sugar into a teacup, while her medical man made various untruthful assertions as to its composition. Bernheim and Liégeois assert that the subject accepted these absurd statements as true because, being hypnotised, she was unable to distinguish between truth and falsehood, while Delboeuf claims that she had sufficient sense left to know exactly what she was doing. To none of them does it seem to have occurred to ask the subject *during hypnosis* what she thought about the matter herself. If this had been done, she would promptly have solved the difficulty by stating that she recognised the experimental nature of the whole performance. It may be noticed in passing that, while Bernheim considers the Salpêtrière subjects so abnormally acute that they can catch the slightest indication of the thoughts of the operator, and so destroy the supposed value of the phenomena alleged to be induced by metals, magnets, drugs in sealed tubes, &c., he, on the other hand, supposes the Nancy subjects to be so abnormally devoid of all intelligence as to be unable to understand when a palpable farce is played before them.

It is also worthy of note that while Bernheim believes in the possibility of suggested crime, he and other members of the Nancy School assert that no actual ill-effects have ever followed the medical use of hypnotism. Thus, Professor Forel says: "Liébeault, Bernheim, Wetterstrand, van Eeden, de Jong, Moll, I myself, and the other followers of the Nancy School, declare categorically that although we have seen many thousands of hypnotised persons, we have never observed a single case of mental or bodily harm caused by hypnosis; but, on the contrary, have seen many cases of illness relieved or cured by it." This statement I can fully endorse, as I have never seen

an unpleasant symptom, even of the most
trivial nature, follow the skilled induction of
hypnosis.

In the " Revue Médicale de l'Est," February
1st, 1895, Bernheim records the only case, as
far as I know, in which death followed hypnosis
induced by a medical man.   The patient
suffered from phlebitis, accompanied by severe
pain ; and to relieve this, Bernheim hypnotised
him.   He died two hours afterwards, and
post-mortem examination showed that death
was due to embolism of the pulmonary artery.
The case is referred to in the *British Medical
Journal*, and though it is admitted that the
occurrence was nothing more than an " unlucky
coincidence," it is stated, at the same time,
that " it is at least arguable that the psychical
excitement induced by the hypnotising process
may have caused a disturbance in the circulatory system, which had a share in bringing
about   the   catastrophe."   Bernheim   has
hypnotised   over   10,000   hospital   patients ;
sometimes this would be for the relief of
pain associated with inevitably fatal maladies ;
and, therefore, the matter for surprise is that
death has not frequently occurred during, or

shortly after, the induction of hypnosis. The majority of fatal illnesses receive medical treatment; it would then, according to the theory of the *British Medical Journal*, be justifiable to argue that the administration of drugs "may have had a share in bringing about the catastrophe." Certainly their use is likely to be attended with more physical and psychical excitement than is involved in the hypnotising processes in vogue at Nancy.

Such arguments against hypnotism are dangerous and apt to provoke unpleasant replies. For example, Moll, in reference to some hostile criticisms of Ziemssen, said: " If Ziemssen had shown the same scepticism when the tuberculine craze excited all Germany, much injury to science and to his patients would have been prevented. The wantonness with which at that time the lives of many were staked will remain as a lasting blemish upon science ; and it cannot be denied that the excessive use of tuberculine was the cause of the untimely death of many human beings. In ordinary life, one would describe such a proceeding as an offence against the

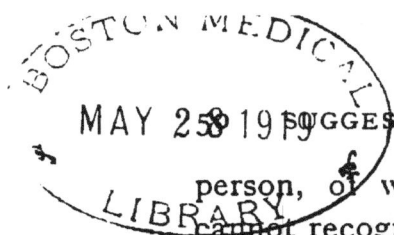

person, of which the issue was death. I cannot recognise that there is a peculiar law for clinical professors; and that when they have, in such a manner, hastened the death of a human being, another expression should be used."

When I commenced hypnotic work some seven years ago, I believed that the hypnotic subject was entirely at the mercy of the operator. I was soon aroused from this dream, however, not by the result of experiments made to test the condition, but from the constantly recurring facts which spontaneously arose in opposition to my pre-conceived theories. Of these facts the following case is an illustration :—

Miss C., aged 19, an uneducated girl, had been frequently hypnotised and was a good somnambule. She suffered greatly from her teeth, of which she had only sixteen left, all decayed. These were extracted at Leeds in the presence of some sixty medical men and dental surgeons. Anæsthesia was produced by written suggestion while I remained in another room. The operation was perfectly successful and unattended by pain, either then or after-

wards. At a later date I examined her mouth
and found that a fragment of one of the
stumps remained. I told her to come to
my house to have it removed. She mentioned
this to one of her neighbours, an old woman,
who advised her to have no more teeth
extracted, as this would cause her mouth to
fall in. The following day she presented
herself and was at once hypnotised, but
refused to open her mouth, or to permit me
to extract the tooth. Emphatic suggestion
continued for half an hour produced no effect.
This was the first occasion on which she had
rejected a suggestion. I then awoke her and
asked why she had refused to have the
tooth extracted. She told me what her
neighbour had said and expressed her deter-
mination to have nothing more done. I
explained the absurdity of this, and pointed
out that, as she had only the fragment of one
tooth remaining, its removal could not affect
the appearance of her face. As she was still
obstinate, I said : " Unless this fragment is
removed you cannot have your artificial teeth
fitted." This argument was sufficient. She
gave her consent in the waking state, was at

once hypnotised and operated on without pain.

This case may be taken as typical of the spontaneous resistance to suggestion in hypnosis, which I encountered. Many similar ones were met with in which the patients refused to perform acts in any way distasteful to them, and that despite the fact that in every instance deepest somnambulism had been induced.

In most of the cases referred to, the patients refused to carry out suggestions in hypnosis which they would have rejected in the waking state. Sometimes, however, they refused in hypnosis things they would readily have done or submitted to when awake. For example, Dr. Allden, when Resident Physician at the Brompton Hospital, hypnotised a girl suffering from chronic pulmonary disease. She rapidly became a good somnambule. On one occasion, after hypnotising her, the nurse reminded him that it was his day for examining the patient's chest ; but to his astonishment, she, although naturally docile and obliging, refused to allow it to be bared. She had previously been examined dozens of times by himself and others, and had never made the slightest objection. He insisted upon her submitting, but was unable to overcome her resistance. He asked her why she objected now, when she had never done so previously. She replied : "You never before tried to examine my chest when I was asleep." On awaking she remembered nothing of what had occurred, and he said nothing to her about it. Afterwards the nurse told her, whereupon she was greatly distressed ; and wondered how she could possibly have been so rude to the doctor.

In the cases above recorded, although a certain amount of evidence was obtained from the patients themselves in reference to their mental condition, no systematic attempt was made to investigate this. I am well aware that this admission is a startling one. I can only say in self-excuse that this all-important point has apparently been equally neglected by others. I have recently attempted to repair the mistake and with interesting results. For example, Miss H., a patient who was the subject of some successful experiments in reference to the appreciation of time in hypnosis, also refused certain suggestions.

I will now draw your attention to one example of this, as well as to her own description of her mental and physical condition during hypnosis. I suggested that she should steal a watch belonging to Mr. Barkworth, a friend who assisted at the experiments. The watch was placed upon the table and Mr. Barkworth hid behind a screen. I told the patient that Mr. Barkworth had gone and had left his watch ; that he was very absent-minded, would never remember where he had left it, would never miss it, &c., suggested that she should take it, that no one would ever know, &c. I awoke the patient. She took no notice of the watch. I asked her, " Where is Mr. Barkworth ? " " Gone away." " He has left his watch, would you not like to take it ? " The patient laughed and said : " No, of course not." I re-hypnotised her and asked : " What did I suggest to you a little while ago when you were asleep ? " " That I should steal Mr. Barkworth's watch, that he was absent-minded, would never miss it, &c." " Then why did you not do so ? " " Because I did not want to." " Was it because you were afraid of being found out ? "

" No, not at all, but because I knew it would be wrong."

On another occasion I again questioned her in hypnosis in reference to this suggested theft. I said : " Did you recognise that it was an experiment ? " " Yes, perfectly." " How did you know it was ? " " I can't tell you, I only felt sure it was." On being questioned further, she said : " Well, I knew you would never really ask me to do anything wrong." " If you were quite certain in your own mind that it was only an experiment, why did you not carry it out ? " " Because I did not wish to do what was wrong, even in jest." " If, then, I asked you to put a lump of sugar in someone's tea and told you it was arsenic, would you do so ? " She replied : " I would not take a watch, even if I knew the suggestion was made as an experiment, because this would be pretending to commit a crime. I would, however, put a piece of sugar into a friend's teacup, if I were sure it was sugar, even though someone said it was arsenic, because then I should not be the one who was pretending to commit a crime." So subtle a distinction would not, I think, have occurred to the subject in the waking condition. In reply to further questions in hypnosis, she said she felt sure she could refuse any suggestion ; that she felt she was herself; that she knew where she was and what she was doing. " Are you the same person asleep as awake ? " I asked. " Yes," she replied, with a laugh.

Strangely enough, the most marked case of resistance to suggestion I have observed was that shown by Liébeault's celebrated somnambule, Camille. When I first visited Nancy Dr. Liébeault showed me this subject, who had been frequently hypnotised, and whom he regarded as a typical specimen of profound somnambulism, illustrating hypnotic automatism in its highest degree.   He assured me that the suggestions he made to her were carried out with the fatality of a falling stone. He hypnotised her, and suggested that on awaking she should find on opening the outer door that there was a violent snow-storm, that she should at once return, complain of this, and proceed to the stove to warm herself. While so doing one of her hands would touch the stove and she would believe she had burnt it.   It was a warm summer's day, and of course the stove had not been lighted. The patient refused to accept the suggestion. Dr. Liébeault insisted for some time, and then gave up the attempt, saying that she sometimes refused suggestions. He then asked her: " Will you do this another time if you will not do it to-day ? " She replied, " Yes, to-morrow." On the following day the suggestion was repeated and carried out in all its details. In this instance, then, the classic hypnotic automaton, the one who was supposed to carry out a suggestion with the fatality of a falling stone, refused one, not on moral grounds, but apparently from pure caprice.

The difference between the hypnotised and the normal subject, is to be found not so much in conduct as in the increased mental and physical powers of the former. Any changes in the moral sense, I have noticed, have invariably been in favour of the hypnotised subject. As regards obedience to suggestion, there is apparently little to choose between the two. A hypnotised subject, who has acquired the power of manifesting various physical and mental phenomena, will do so in response to suggestion, for much the same reasons as one in the normal condition. In the normal state we are usually pleased to show off our various gifts and attainments, more especially if we think they are superior to those of others; and in this respect the hypnotised subject does not differ. Both will refuse what is disagreeable; in both this refusal may be modified or overcome by appeals to the reason, or to the usual motives which influence conduct. When the act demanded is contrary to the moral sense, it is usually refused by the normal subject, and invariably by the hypnotised one. I have never observed any decrease of intelligence in hypnosis; in the alert stage it is often conspicuously increased, while in the lethargic it is only apparently, not really, suspended.

When one turns to the later works of Braid, and sees how clearly he proved by experiment that the hypnotised subject not only had the power of choosing between suggestions, but invariably refused those repugnant to his moral nature, one cannot help feeling surprised at the revival of theories in reference to so-called automatism or obedience, identical with

the views of the mesmerists. More especially
so when one considers that Bernheim, who
holds these views, also boldly asserts that
there is nothing in hypnotism but the name;
that it does not create a new condition, and
that hypnotic acts are only exaggerated
normal ones. According to Bernheim, however, the moral state in hypnosis differs widely
from the normal; and this is in obvious con-
tradiction to his own conception of hypnotism.
One can understand, for example, how a
prolonged muscular rigidity may be a hypnotic
exaggeration of a somewhat shorter normal
one; but it is difficult to comprehend how
the murder of one's mother when hypnotised
can be an exaggeration of the *refusal* to hurt
a fly when awake.

It is now seven years since I commenced my
hypnotic researches, and during that time I
have not seen a single instance, either at home
or abroad, in which the subject has performed
any real act which was distasteful to him
or repugnant to his moral sense. While
incredulous as to the possibility of suggesting
criminal or improper acts, I still think it
important that hypnosis should be induced in

such a way as to render interference with the volition of the subject impossible. The patient should be made to realise that the phenomena of hypnosis are entirely due to the exercise of powers existing in his own mind and under his own control.

In illustration of my theory that the phenomena of hypnosis are not necessarily connected with any interference with the volition of the subject, I desire to draw your attention to two classes of cases :—

1. Where the operator has deliberately tried to minimise his own importance in reference to the induction of hypnotic phenomena.

Although I soon ceased to believe that the subject's volition was dominated by that of the operator, I still found, as the result of sensational writings on the question, that a considerable number of my patients objected to be hypnotised, on the ground that it would interfere with their volition. To obviate this difficulty, I changed my method of inducing and managing the hypnotic state. I com- menced by informing every new patient that I did not believe it possible for the operator to dominate the volition of the subject, and that, even if such a thing were possible, it could certainly be prevented by suggestion. I explained to my patients that nothing would be suggested without their consent having been previously obtained in the normal state. Under these circumstances, if the suggestions were successful, this would not imply any interference with volition, seeing that their consent had already been obtained. I pointed out that the fulfilment of a hypnotic suggestion frequently demonstrated an increased, not diminished, power of volition. For example, a patient who desired to resist a morbid impulse, but was unable to do so by the exercise of his normal volition, might gain this power by hypnotic suggestion. Thus, the suggestion did not suspend the volition of the subject, but removed the obstacle which prevented the wish being carried into action. Further, as resistance was manifested despite suggested obedience, it was reasonable to expect that this might be enormously increased by training. I suggested, therefore, to all patients during hypnosis, that they should invariably possess this power of resistance, and also that neither I nor anyone else should ever be able to re-induce hypnosis without their express consent. This change of method did not affect the results. Notwithstanding the fact that the patients were convinced, and justly so, that they possessed complete control of the whole condition, hypnosis was evoked as easily as formerly, and as wide a range of phenomena was induced.

2. Where an attempt has been made to teach the subject to evoke hypnosis and its phenomena without the intervention of the operator.

Some six years ago I commenced to instruct patients how to hypnotise themselves. This was done by suggesting in hypnosis that they should be able to re-induce the state at a given signal ; as for example, by counting " one, two, three." These subjects could afterwards evoke the condition at will. I also found that the use of suggestion during hypnosis was not necessary for the induction of its phenomena. On the contrary, the suggestions could be made equally well beforehand in the waking state. The subject was able to suggest to himself when hypnosis should appear and terminate, and also the phenomena which he wished to obtain during and after it. This training was at first a limited one; the patients, for example, were instructed how to get sleep at night, or relief from pain. They did not, however, always confine themselves to my suggestions, but originated others, and widely varying ones, regarding their health, comfort, or work. Some, trained in this way six years ago, still retain the power of hypnotising themselves.

In such cases it would be difficult, I think, to explain hypnotic phenomena as the result of arrested, or weakened volition; or of outside interference by the operator. It might be objected, perhaps, that the influence of the operator had not been entirely eliminated, on the ground that he had been associated with the induction of the primary hypnosis. The conditions, however, which are more or less frequently associated with the origin of a particular state are by no means essential for its after-manifestation. For instance, the art of swimming is usually taught either by means of a life-belt, or by attaching the pupil to a cord, which the teacher holds and guides by means of a rod. These artificial aids, however, are not essential to the art of swimming ; they are only useful in its acquirement. It would be illogical to ascribe a champion's power of winning a race to the presence of a life-belt which he discarded years before. In the same way, it would be unjustifiable to attribute a subject's power of influencing forces within his own body, by suggestions rising in his own mind, to the influence of the operator who had formerly instructed him how to evoke and direct this power.

Owing to their number and diversity, the sketch of hypnotic theories I have attempted to give is necessarily a very imperfect one. Max Dessoir, in his " Bibliography of Modern Hypnotism," published in 1888, cites 800 works by nearly 500 authors, and since that date many more have been written, while it would be difficult to find two authorities agreeing in every detail as to the theoretical explanation of all hypnotic phenomena.

During the last twelve years the practice of hypnotism has spread over nearly the entire Continent of Europe. It is now not only largely employed by the ordinary medical man, but is also to be seen in daily use in hospital wards, while many distinguished physiologists and psychologists have published their observations of its phenomena. Apart from Nancy, the two most remarkable hypnotic Cliniques I have visited are those of Amsterdam and Stockholm. The former is under the direction of Drs. van Eeden and van Renterghem, who have published two accounts of their work giving the history of 1,089 cases treated from 1887 to 1893. The latter is conducted by Dr. Wetterstrand, who up to January, 1893, had treated 6,500 cases, and failed to induce hypnosis in 3 per cent. alone. Many of his clinical observations are of great value; one of the most interesting parts of his work is his treatment of epilepsy and various other forms of grave nervous disease by prolonged sleep. The patients rest quietly in bed in a condition of hypnotic trance, for periods varying from a week to a month, or even more, and meanwhile are fed at regular intervals by nurses who are put *en rapport* with them. This method, which adds mental rest to the ordinary " physical rest cure," I have myself successfully employed.

In addition to those to whom I have already referred, the following, amongst others, may be quoted as having recently testified to the therapeutic value of hypnotic treatment :— Forel, Freud, Gerster, Grossmann, de Jong, Scholz, von Schrenck - Notzing, Tatzel, Brunnberg, Hecker, Krafft-Ebing, Ringier, Bergmann, Brügelmann, Fulda, Herzberg, Hirt, Schmidt, Vogt, Schmeltz, Lemoine, Joire, Voisin, de Mézeray, Bérillon, Dumont-pallier, Gorodichze, Bonjour, Desplats, Bourdon, Tissié, &c. The nature of the cases treated varies widely, the most common, however, being different forms of functional nervous disease.

Possibly the most interesting and striking instances of the value of hypnotic treatment are to be found in the cure of morphinomania, dipsomania, and other moral diseases. In the " Zeitschrift für Hypnotismus," Part I., 1896, Wetterstrand reported 38 cases of morphinism treated by hypnotic suggestion. Of these, 28 were cured, 3 relapsed, and in 7 no result was obtained. In each instance the morphia had been injected subcutaneously. Many of the cases were exceedingly grave

and of long standing, and some were
complicated with the cocaine and alcohol
habit. With several the abstinence treatment
had been tried without success—sometimes
more than once. One of the successful cases
—a medical man—had taken morphia for
eighteen years, and during the last four years
cocaine also. Another medical man, Dr.
Landgren, recorded his own case in the same
journal. Over five years have elapsed since
he was successfully treated by Wetterstrand.
Other methods, including residence in a
retreat, &c., had failed. Other observers have
reported similar cases, and in many instances
sufficient time has now elapsed to enable us
to judge of the permanency of the cure.
Some of those with whom I am acquainted,
who formerly suffered from dipsomania, are
now actively engaged in business, or in
successfully conducting medical practice;
one has since been elected a Member of
Parliament, while others are happy wives and
mothers. In most of them the disease had
been of long duration, varying from about five
to fifteen years, and in some presented all its
worst symptoms ; thus, one patient had had

several attacks of delirium tremens and
epilepsy.   The duration of the cure has lasted
from two to over six years.

I do not for a moment pretend that by
hypnotism one can cure everything and everybody, and agree with Braid in thinking that
he who talks of a universal remedy is either
a knave or a fool.   On the other hand, I have
seen many cases cured or relieved by hypnotism which had previously resisted long,
careful and varied medical treatment.

In this country those who practice hypnotism
have to contend with greater difficulties than
are to be met with at Nancy.   There, as the
result of Liébeault's thirty years' work, fear
and prejudice have been almost entirely removed, while here they not only exist, but are
widely spread.   For this, disgusting music-
hall exhibitions, generally of a fraudulent
nature, are largely to blame, although
undoubtedly the erroneous views as to interference with volition, &c., expressed in the
British Medical Journal have done still greater
harm.   Hence, in England, patients rarely
turn to hypnotism except as a last resource
and when all other methods of treatment

have long been tried in vain. Despite this, the results are encouraging, and have in many instances withstood the lapse of time.

One point is worthy of special notice. The results obtained by hypnotism are due to the action of the patient's mind on his own body, and are free from the disadvantages frequently associated with the use of drugs, or other similar remedial agents. Thus, the patient who is cured of insomnia by hypnotic suggestion has really acquired the power of producing sleep at will ; while he who obtains sleep by means of an ordinary narcotic has often simply added a drug habit to his original disease.

My personal hypnotic work has been almost entirely of a therapeutic character; but hypnotic research, both physiological and psychological, has undoubtedly a wide field open to it. In reference to this a word of warning is perhaps not misplaced. Experiments should only be conducted on those in health, and never without express consent having been given in the waking state. Patients, above all others, should be regarded as sacred, and nothing should be suggested to them,

except what is necessary for the relief or cure
of disease.

Now that the reality of the facts of hypnotism
have been so abundantly proved, and successful
simulation rendered impossible by the production of objective phenomena which no one can
simulate, further scepticism and indifference
seems scarcely conceivable. How be indifferent
to a science which explains so much of what
is mysterious in history, which supplies the
psychologist with a genuine experimental
method, and furnishes such aid to the study of
cerebral physiology ?   How neglect an art by
means of which pain of mind and body can
be relieved, and a large number of diseases
cured ?   The historian finds the whole subject
of Magic, Oracles, Sybils and the like illumi-
nated with a vivid light.   The miraculous
character of religious ecstasies, apparitions
and stigmata disappears, and to Virchow's
dilemma, formulated in 1875, concerning the
celebrated Louise Lateau, " Either trickery
or miracle," we may triumphantly reply,
" Neither trickery nor miracle, but *Suggestion.*"

To the psychologist, a harmless and painless
means of intellectual and moral vivisection is

given ; while the physiologist, by the aid of hypnotic suggestion, can produce anæsthesia and other sensorial conditions, and regulate or modify the secretions, excretions and other functions of the human body, which the waking will is incapable of influencing.

While those who study and practise hypnotism differ widely in their theoretical explanations of its phenomena, what is common to all, as Professor Beaunis states, is the belief in its future—the profound conviction that this science, so ridiculed, constitutes one of the greatest advances of the human mind and one of its most precious possessions.